I HEARD SOMETHING

I HEARD SOMETHING
JAIME FORSYTHE

AN IMPRINT OF ANVIL PRESS

Copyright © 2018 by Jaime Forsythe

"a feed dog book" for Anvil Press

Imprint editor: Stuart Ross
Cover design: Rayola Graphics
Interior design & typesetting: Stuart Ross
feed dog logo: Catrina Longmuir
Author photo: Alvero Wiggins

Library and Archives Canada Cataloguing in Publication

Forsythe, Jaime, 1980-, author
    I heard something / Jaime Forsythe. -- First edition.

Poems.
ISBN 978-1-77214-123-8 (softcover)

    I. Title.

PS8611.O77277I34 2018   C811'.6   C2018-901890-9

Represented in Canada by Publishers Group Canada
Distributed by Raincoast Books

The publisher gratefully acknowledges the financial assistance of the Canada Council for the Arts, the Canada Book Fund, and the Province of Columbia through the BC Arts Council and the Book Publishing Tax Credit.

CONTENTS

ONE

    Constitutional / 11
    Notes for the Housesitter / 14
    Autobiography I / 15
    Introduction to Art Heists / 16
    Who Is Missing, Who Is Missed / 18
    I Heard Something / 19
    Instructions for Heavy Weather / 20
    Preparing to Live / 21
    Island / 22
    Autobiography II / 23
    Beluga / 24
    August First / 25
    What's Unsolved as a Place to Rest / 26
    Our Experiences Were True / 28
    Last Night on Earth / 29
    Vacation Rhinoceros / 30
    Inside / 32
    This Isn't Me / 34
    Once or Many Times More / 35
    Late Admissions / 36
    You Know / 37
    Your Tour Guide Is a Recording / 38
    Wrong Number / 39

TWO

    Museum of Natural History / 43
    Skin to Fish / 46
    The Housepainters / 47

The Hurricane Simulator Is Under Construction / 48
A Good Candidate for Space Camp / 50
Nothing's Wrong / 51
Current Mood / 52
Strong Suit / 53
Roles & Responsibilities / 54
Archetypal Story / 55
Did We Blow a Fuse, Love? / 57
Ritual for Getting Rid Of / 60
Using a Part of Your Body, Write a Letter in the Air / 62
Interview Before Departure / 64
Force Field / 65
Little Proofs / 67

THREE

If Lightning Was a Thing You Could Plug Into / 75
Spells / 76
Interpretation of Symbols after Months Without Sleep / 77
Play Police / 79
Dear Stranger / 81
You Be a Ghost / 87
I Lie on the Floor in the Dark / 88
Sewing Machine / 90
Mom's Night Out / 91
Autobiography III / 92
Kinship / 93

...if
I open my eyes I will know the room very well
there will be the little thrown-out globe of blood we left
and every molecule of every object here will swell
with life. And someone will be at the door.

—Jean Valentine, "The Knife"

ONE

CONSTITUTIONAL

Plane balanced on a cloud
carves a grave telegram.
If we gather our venom in one place,
it will die in the fire. Tumour be gone.
Foaming under flannel, long
for weather, long for nuclear
feeling. A robber poached
the change jar, left our linen
floating on the line.

One night only, a memory screens
on a farmhouse ceiling: the stars
onstage ate sponge cake
before playing. We held our jealousy
careful as injured wrens.
Swampy vocals, dirty fork on an amp,
smashed lilies in the circle pit
made a map of the downtown core.
Roped into an espresso at midnight,
strange coos in the alcoholic
hallway. The only one at the punk show
in a peacoat. When shown the sky,
a new device pinpointed Ursa Major. Snares
shivered lonely on the concrete.

Never too early, never solo.
Time melts as mint concoctions stream
from the shaker printed with remedies.
Suspended on our backs, stuffed, ear

canals filling with brook,
the patterns on the bridge's underside
a silent film. Air thick with the bee
mortality rate, crops kept
from curing arthritis, from leaking ink
for acid lyrics. Plastic hot dog
lost in the current.

Rawhide in my pocket
trails a ribbon of scent, perks up
the dozing huskies. Leafy swish
of her crinoline, faceless
monsters in the corn.
Please respect the corn.
It took us months to grow it,
and your parents first met
in its rustle. Just when I find solace
against a cool wall, a panel begins to move.
The blue room is bare,
but if you stand in its heart,
you can feel the weight of smock,
stethoscope, scalpel, energy
of the current that keeps your spirits high.
Fog is a curtain around the operating table.

We tried to coax the bees into the squash
flowers, but the seeds were rotten to start.
They pour something blue on the beds
and enjoy pastel drinks laced with fertilizer.
Dessert ruined: cream shook to butter
too quick, conversation a mudslide.

The rhythm is recorded first before
the voices, one at a time. At one time, women
banded together: cigar makers, tailoresses, umbrella
sewers, cap makers, laundresses, shoe workers.
A felt-skirt factory, a picture-frame factory.
Many roamed the cities looking for food.
Their babies took what they had
and placed it back into the ground,
eating earth, drooling
as their teeth grew in.
The little ones squinted at shrubs
and dreamed of woolly mammoths
chewing at their land.

NOTES FOR THE HOUSESITTER

The china cow who spits cream and the commemorative Queen Elizabeth II plate are well-known to one another: no need to introduce them. Under the mudroom sink, find a basketful of multicoloured ropes. These are for athletic use only. You do not have to make the spices form a conga line of ascending heights, no matter the occasion, or caress the smooth leek in the bowels of the crisper if this is not appealing. You do not have to follow silver arrows made of electrical tape, or consider every cult offer that comes knocking. Please bring in the mail.

The mints on the dresser are all yours. I always knew you were a cubic crystal, octahedral, shiny with a burnished lustre. Ignore the playing cards pinned to the bedroom wall in the sequence of clubs, diamonds, clubs; they represent an internal mystery I've yet to decode. If the fish dies, cart him down the path and drone a benediction before his final liberation off the dock.

Trim the cacti weekly—as a personal favour to me. A muskrat standing at the back door foreshadows a sudden relative. When going out, it is your job to skim backwards in violet dirt, minnow your way past the toll. Finding a return route is the eternal conundrum. Do protect your parts and belongings.

You do not have to fill a gorge with everything you have. You do not have to start again from the beginning, shoving a metal detector to uncover your own ancient footprints, brandishing glue and dull scissors. In an ideal world, I return and find you still in one piece.

AUTOBIOGRAPHY I

I paused, but didn't expect anything else to. Maybe others would join me on my island, maybe not. I missed that city, its neon tetras through glass and haunted corners by closed laundromats, but when I returned, it had been emptied of its former associations. A paper bag, the air pressed out of it, that I could fold and stick down my dress. The crowd was so slow because each hand had to touch the hand on its left before advancing through the turnstile, down the ramp, and past the dancing ribbons. A fringe of fingers played air piano over a used car lot. The intersection where I was once yanked into a taxi's padded room pulsed red. The streets had been dipped in green but still I flinched like an inchworm, wanted to tuck myself into a plush music box. While on a locked ward, I received three marriage proposals: I had never been so popular and haven't been since. A pink envelope came in the mail, and the floral card said *If You Don't Relax I'll Kill You*. I started to understand what it is people want. Most days, I could be satisfied fiddling with cinched telescopes and safety-pinned waistbands. My head sank into the hands of the person I'd hired.

INTRODUCTION TO ART HEISTS

When I drew closer to the reclining centaur,
its breath hit me, salty in the cold, clean place
where I decided to commit the most serious

crime of my short life thus far. It seemed fair
to touch the plaster of Paris, to stroke the fur
and trace the embroidery, decisions made

easy next to the rich dilemmas
I'd been struggling with. It was a matter
of acting natural. I got lost in the tease

of details: a drip of red on icy floors, a hair caught
in the light switch, a bullet hole in the wall they forgot
to spackle. I paused at a projection of clouds, faces

turning into clouds. The sculpture was on
my side. I spotted my doppelgänger and changed
directions, fixed the angle of my elbows;

a small adjustment is all it takes to get the safe
combination right. It was only your hands
shaking the first time, jarred by the half-animal

hiss of last decade's walkie-talkies. Security
was baffled enough to lag, and the polished shoulder
fit under my arm like a dream, like the dream

I had last night, where parts of the city were missing, and my clothes hung in the trees, and the street lights dissolved one by one, all the way home.

WHO IS MISSING, WHO IS MISSED

who holds the sign, who knows their SIN, who is inoculated, who is hooked up, who is prone, who is photocopying their palm, who is photographing themselves to check if they are still visible, who brought sandwiches, who broke the piñata, who got carried away, who had been drinking, who lucked out, whose attention was halved and quartered, who turned up the wrong burner, who is taking minutes, who made the rubric, who is psychic, who is engaged in scrupulous self-care, who can swim, who has insurance, who slept with whom, who made out like bandits, who missed their flight, who dumped the evidence, who was asked to weigh in, who spoke to them last, who was followed, who was the messenger, who packed this luggage, who lives here, who is on the list, who listened closely, who never heard of it, who graduated from the institution, who let the door slam, who knows the password, who is still up, who wants tea, who should be notified, who made the sign, who didn't see it, who swallowed their key, who was thanked, who has a good back, who made squares, who is bilingual, who is overdue, who is welcome, who wants to live, who shows promise, who is whole, wilful, howling, well?

I HEARD SOMETHING

Upon inspection, what made the sound was not obvious. My days of sleeping soundly were over. Unsettled, I struggled to locate its source, eliminating woodpecker, city bus, coffee grinder, dying television, insect in pain, cello. It did not bring to mind anything that might be emitted from the folds of a human, animal, or plant. It did not disappear into any identifiable music note, in or out of tune. Once in the early days I followed its frayed thread as far as I could, and found myself crawling through a concrete tunnel, then standing in a baseball diamond at midnight. Of course I wondered if it was coming from me. I wondered that right away.

I wished for soundless dreams. I wished for an aural microscope. Headphones, earplugs, blindfolds—none made a difference. When an audiometer gave me a perfect score, I went weak with the betrayal. The sound rose from the grass, or the warm skin of whoever was asleep on the floor beside me. I wanted to bury myself in the earth or other bodies, enter it, snuff it out. If it had a form, I would have welcomed the opportunity to study it, tip it from side to side in a glass vial stopped with cork. No dog seemed to detect it; in fact, many species dozed uselessly through its most robust moments.

How could something so slippery so rarely leave me? I devoted myself to living with it in peace, like a pet gerbil or benign mole, however strange it might seem to attend to a vibration that exists only privately. Other people, though, must have their own incarnations. I had company, there was that. It's true I was no longer alone.

INSTRUCTIONS FOR HEAVY WEATHER

after a collaboration with Alice Burdick

You egg cup, you balloon animal,
shatter and burst, dilute without
fuss. Two celestial bodies nod
hello while a bucket of paint overflows
in the rain. Beach your testimony
for a tried-and-true myth. Fiddlehead
your hair for the ceremony? Not enough.
In the wet glow, ghost your misery.
Freeze the clutch inside the hi-hat's cloud.

Anything can be a hat. Clean and dress
and meet me at the back of my head.
Remove the small sac from my nape,
the one I was saving for a hurricane,
along with the petrified fruit in the cellar.
Wild or cultivated are your choices for supper.
The speeches, the rice, the gaping
sky: a cavity to be closed with stray eyelashes.
Watch the porch light's seizure, silver moths.
Wait for everything to stop.

PREPARING TO LIVE

Rain tap, radio off or on to keep
the foxes at bay. Ceilings screwed
tight, chandeliers lowered, leaf
inserted, films rewound, speakers tricked
into silence. The ways we prepare
the house before leaving it. The way we house
our grains and percolators, our trivia
and instincts. An iceberg eclipsed
the shore, showing off but also not giving
a fuck. A hairpin turn in gravel was crucial,
we felt, to outsmart the deer we were.
Air perfumed with notes of ozone
and chloroform, he spooked and ran,
but not before giving us one good look.

ISLAND

The thought machine is dismantled by an escaping heron. A pelican's dive looks like a mistake, but we've been separated from our pod, know nothing but this conveyor belt of sand. Spaniels in security vests are lax at the border, letting us in, despite everything. We don't even know which animals are here: are there llamas? Is a leopard tour possible? I can't explain the man who evaporates from a bench, or the woman in bronze makeup floating cross-legged above a hibiscus, or the sequined headpieces gliding into the pool as Céline Dion ribbons out from a tape deck. Horned shells go spiny and indignant in our pockets, leaving bite marks on our veined thighs. We press anxiously against perfect shots: a cactus, a cactus, a cactus. The waitress's crooked incisor flashes as she pours Tang. In a circular room, palms grow from the floor and erupt through the ceiling, their tops too far away to see.

AUTOBIOGRAPHY II

I was beamed or flown—it's fuzzy.
I don't remember the first place
I lived, or the second. The third
was a bluish-white dome, a hole-
punched sky, bird cages against
thickly layered wires, power
sucked out of the compound. Then
star anise, hands trading liquids
and papaya. I woke on a hill
flecked with radios reciting
nonsense or code, reception sharpest
at the peak when I balanced there.
I was permitted a leashed wandering.
We walked and sipped, tilting
toward surfaces where we could close
our eyes and mouths, springed or flat,
tune out the scuttle in the walls.
We were vigilant of the margins,
of sparks and wavelengths. A new day
where I resume my hunt for my earthly
counterpart. Pressed temples, a turtle
surge into a holy pool, wobbling across
a planetary body.

BELUGA

You created the conditions that would coax me to you. I felt invited, knew a search for the wrong kind of company, the fine distinction between movie and mirage. I paddled the cove hugged by swaths of cracked bone. All I wanted was deep cold and a catalogue of sounds, the pitches of barbershop singers, espresso machines, and plastic pianos for babies. I thought I'd start a band called Sea Canary, had heard rumours of East Coast luaus and amphibious vehicles. I hoped to be more, profile photo in an inflatable swan, more, repeat visitor at the swim-up bar, more, happy with a handful of almonds for lunch. I blew raspberries into your harbour and waited. Don't swim with it, don't pleasure craft around it, the radio admonished, voices boomeranging off coloured domes that opened and closed, depending on the sky. Little did I know that the longer we remained in proximity, the higher the risk. I look approachable due to my eternal smile.

AUGUST FIRST

A man in swimming trunks
steps out his front door, stitches
himself onto my stride, says he finds
this place unfriendly. The shopping cart stuck
in the parkette won't budge. A mannequin's hat
stutters past the café. Summer of softball mercy
rules, pop cans, fountain spray, naps in the bleachers.
I play dumb in an ergonomic chair, ask questions
I know the answers to. And he's right that
everywhere is deadpan, wary. Yet plots balloon
with tomatoes. Pea tendrils curl tight around whatever's
close. Yellow sky and long lineups at the lost & found,
hot demand for lone mittens to float between innings,
past pennies quiet in their turquoise bath.

WHAT'S UNSOLVED AS A PLACE TO REST

I waded into a warm purple night,
witnessed fur growing around the base

of a telephone pole, but missed
the tossed Molotov cocktail

that made the papers the next morning.
Yes or no! the moonstruck suspect

was quoted. Why does anyone
do anything, ever? Walk into the woods,

get lost in their own pupils
in a dressing room mirror, go home

with a stranger, allow
their worst instincts to ricochet

out from their bodies. What more
flew by me, sunk as I was

in my own fog? A ride drew up,
idling in confused exhaust.

On TV, a lithe PI slipped through
a doggie door without a warrant.

I had to be more like that,
economical, not wait

for permission. I leashed
my dog and we criss-crossed

the neighbourhood, looking
for the long-lost girl

I knew. I didn't know
what I expected

to solve. As a girl
detective, I trained by peeking

at soaps and procedurals through
dollhouse windows, absently

posing the rubber people,
making them make

their tiny breakfasts and sit
in the tub with their clothes still on.

OUR EXPERIENCES WERE TRUE

Washing mouths out
in the river, we developed
a private rash. Undocumented
sounds—cellphones, pipes?—
rattled the town and
moved us to open a hole
in the earth, bewitched by
opulent ideas like buried
treasure, and not unaffected
by preoccupations, voices,
et cetera. Was it clairvoyance,
or just paying attention?
The healing sector had long
ago thrown up their hands,
so our modest spades
seemed worth a shot,
allowed deep dives
into man-made sandbanks
where we found
a claustrophobic ecology
of anemones and buttons.
When we hoisted ourselves
back to crisp reality, spilling,
they were poised
on the beach, pressed
their fingers to our lips.
Shush, they said soundly.
We've already made up our minds.

LAST NIGHT ON EARTH

When you called me a ringer
for someone famous or dead, I got
misty, abandoned my waterproof cape.

A rotating vase purred
in the space between our faces.
By the time the staring contest

drew tears, capsules
evaporated, stained the air
with medicinal pleasantries.

We curled toward
the bed's axis, saw the same
forms descend from the chandelier:

wheels of leaping antelope.
Mints fossilized on our tongues
like little bones in a museum.

VACATION RHINOCEROS

A jewellery box glints with nervous tics.
Ballerina swivels on her stumps
to Greensleeves. It's not easy, finding
the nucleus of the spider bite as it swells—
to hush-hush wounds or wrap them?
Dark-eyed junco, hop in butter light.
We pressed playrecord to archive
our voices, dubbed over a book-on-tape.
We staged a play for our uncles, let
the sun set to Chopsticks on piano.
These are not delicate men, but they
are light. A plastic one with parachute
released from the duplex patio. You
won a balance-beam competition in
your backyard, while a chorus
of paper gowns waited in the wings.
A spoon racket, whiskered platitudes
over toast. You sent me your vacation
rhinoceros. Be aware of the staircases
and padlocks you attach to your rooms,
raccoons multiplying in the night.
It grew damp inside the rubber
horse mask. Maybe stuff a bag
and make a head from it?
The grave was a real letdown,
a flat nothing place.

Every figure eight and bridge pose
and loop de loop joined forces
to dump me here, on this fragrant
and depth-altering field.

INSIDE

A jar gulps rosettes.
Aerophytes sip air
and bathroom mist, thirsty hairy
limbs, while
black-eyed Susans
and cattails plump
a juice container—
we are good
in our response to
sicknesses
that settle.
We tie bows, boil broth,
serve tarts on scalloped plates.

O terrarium
I pet your glass
and remember my
outside times, shifting
dunes, fields erased
by Queen Anne's lace
in August. Atrophied
roots a gateway
to slow drift, a tucked
veined heart, fireworks
across closed eyes.
I cannot hunt—
who would hunt? Supper
a sea bream flown here

on ice. Who has time
to forage? I gather
what has been done.
The cold makes designs
on the window,
my one newsreel—
what trunk, what wheel, what topple.

THIS ISN'T ME

Street chess erupts, hurled rook—
who here wants to go to heaven?

Race to the bedsheet that's not
a person bent into the sidewalk.
Expect nothing for your quivering

antennae. This isn't me—
this isn't the kind of hat
I'd wear normally. Beaded vines

click in the smoky shop
that tastes of a chilled spoon,
like when you're being put under,

falling through layer after layer
of scarves waving like seagrass.

Sorry to bring you here.
Soon you will board a merry gondola.
Soon you will make brilliant moves.
Please lend me a sensation to pack
for the coming voyage.

ONCE OR MANY TIMES MORE

I wear the same clothes over and over. I am not weary, but enjoy being swallowed up in a forgiving way. I gulped down a plum pit, and the trunk that took root in my stomach was surgically removed at the quick. It is easiest to recall what I tried only once, the rituals most quickly discarded. Over and over, I wear people out. I do the same thing over and over, yes, expecting a different result. Or maybe I liked the result the first time, am trying to recapture its original sheen. I pen a score but burn it, prefer to perform my repetitions by heart. My grooves are good, my knees and soles, transparent, breaking in two. I am undone over and over by the daily care of limbs, organs hogging every minute, hair streaking the tub. Over and over, I enter a room of flowers and altars, pews of wet stares, slip my arm through a rope and sway. The same song hovers a beat ahead of time and I pursue its trail, humming. Over and over, I hope to reach a speed where I might escape my own unbuttoning skin. I stomp through the wild in boots, and the branches trampled yesterday have sprung back into place. It all just keeps coming, tickertape, waterfall, scroll. I find a red-winged blackbird dead and don't know what to do. Over and over, I untangle my obligations from the ground. Asterisks twirl down but also reverse cloudward in the candle's flame, doubtful matter, every season they do this, and over and over it stops me cold.

LATE ADMISSIONS

I did not transpose my grief
into those falling clumps of snow.
I did rip my sunburn into white
strips last summer, one whisper closer
to the centre, raw and exposed from soft
cups of keg beer. I told the person
pouring how I missed my positively
altered states and euphoria, post-flu.
They dropped this info into a vacant
drum. I did not see the supermoon
because the uncooked dough
resembled someone I knew.
Can I blame you? I know what
I saw, that an eye twitch
is a dispatch from another dimension
in a tenacious language. Believe me,
I'm listening, just wait.

YOU KNOW

I can't tell you why a dalmation
is growing in this dirt. I only planted

mesclun and beans and nothing took.
I got over it. The landlord pressure

washed the mess off the side of the house,
told me he lost half his sight

from a lawn dart, once floated his glass eye
in milk to scare the babysitter.

I squeeze into my bathing suit, arrange my eggs
in their holders, prop the refrigerator open

for cheap refreshment. It's too quiet
and I can't stop listening to songs that go

Ooo, can't stop my useless premonitions. I knew
the long-distance ring would trill, the lump

would have nothing in it, that you wouldn't ask.
Anxiety zings through me, attacks

my copper manicure. To gather courage,
I practice-call on a zucchini phone.

Maybe pick up something while you're out.
You know what I like.

YOUR TOUR GUIDE IS A RECORDING

Since last passport photo, less ability to arabesque
for the public. Spike in cravings for startling
architecture, chilled dragonfruit
on a dry throat. Fear, though, of faulty engines
and small vessels spinning in peaks
of whipped snow. How sealed
must a pair of palms be
to carry a living thing from one perch to another?

It's healthy to have visions
of stirred morning glory, glowing
lowercase letters poured over concrete, tape
tuned to the language that best suits your genetic
tongue curl. Psalms flow into the canal. Steps from the street,
a bombed vineyard grows
from its fragments, and shoes point to Saturn

frozen in tile. Politics get blasted legibly
on dumpsters as you agree to meet by the headless bust
in front of the government building. Dogs tied to iron
bark and lap spilled latte, shiver in the courtyard,
headphones murmuring *a hidden passion*.

A lion-head doorknob. The virgin and vines
suspended out a window, and laundry,
a passing urge to be left
in a well, yet something
is keeping you here.

WRONG NUMBER

wrong shade of blue   wrong time of the month
    wrong part of the lobster   wrong party trick   wrong
glitter bomb   wrong prayer   wrong emoji
    wrong perfect little hole-in-the-wall   wrong measure
of emotional heft   wrong spell   wrong species
    wrong era of vintage dress   wrong symbolic gesture
wrong definition of intimacy   wrong tchotchkes   wrong
    choice of dentist   wrong balance of light and dark
wrong line of work   wrong rhyming couplet   wrong spasm
    wrong toxin   wrong degree of disclosure   wrong treasured
influences   wrong terrestrial planet   wrong history
    wrong birthright   wrong birthmark   wrong dark alley
wrong recovery time   wrong boundary   wrong thing to say
    afterwards   wrong abyss   wrong buzzer
wrong neural pathway   wrong path   wrong hair removal
    technique   wrong rodent trap   wrong fizzy comfort
wrong cleft   wrong entryway   wrong tonic   wrong tone
    wrong code to transmit intended

        desire

TWO

MUSEUM OF NATURAL HISTORY

Milky after-school light. Pines pressed into the backdrop,
draining colour from the sky that hugged their jagged

outlines. Bald eagles in flight, over a painted canyon
the illusion of deep. Black bear's growl looped

through a speaker. Slow pitch forward when faced
with a revolving model Earth, misplaced gravity,

a whale to scale, drifting near the ceiling—
all we could see from where we stood, a string

of children on a wooded path, paused in a shadow
to listen to the moss. Burrowing into warm

chlorine, unnatural breath, floating
toward vibration and blue. Theatre

of backlit fossils bloomed from rock,
blessed by our fibre-optic wands.

If we held our positions long enough,
dinosaur skeletons filed from the building

to chew trees and divert traffic, indoor forest
with sound effects, where footsteps left behind

no mark. Penises chalked in a sloppy series,
lines wilting in an unwatched classroom. Curses

gorgeous on brick, baking in the sun. Felt pennants
thumbtacked to the neighbour's green bedroom,

eyes of wildlife ornaments and trophies moving
in sync. Chenille kneaded in a pincer grip, perfected

at six months old, while parents' tumblers collided
in the study. A tent of butterflies that swirled

and hitched to our sleeves and hair, lending
their wings for a single flap. Heat that kicked in

from ancient vents and stirred fledgling allergies,
sifting the ruins of postcards: pillars, temples,

plazas, columns, dust. Waking up to vortexes,
waking up to blood, portals unlatching in chest

cavities. A hamster crept inside a transparent
globe, sightseeing a braided rug. A hand on a knee,

a hand on a hem, a hand on a chin. The difficulty
of finding a way to leave a room, let alone a tower, let

alone a town. A door pushed open, a fold lifted
and smoothed. Skits cobbled from tablecloths,

keyboards, paper crowns, the kind of kids
who'd repeat anything. Reaching for a script

whenever walls began to move, coming up with only
Mad Libs, blanks mocking. Light spoked over organ

pipes, striping the restless choir. The mammals, the insects,
the plants, the coral, the skate behind thick glass:

all dead. The voices, swelling until we stepped
back. Gravestone rubbings done with crayons,

hoods nodding in the breeze. Time measured
with a ribbon, then a ruler. Perpetual calendar,

pencilled numbers, our small, strange
powers. Salt crystals raised in cupboard

Mason jars, gems thriving in dim caves.
Velocity of a liquid train, drink shaking

on a tray, and out the window
lit maps that blinked and beckoned,

true only while moving:
you are here, you are here, you are here.

SKIN TO FISH

Palm skims the length of a halved salmon,
hands cleaner than any glove. Plates
lumped with pods and shells,
cast-offs from faraway
life forms and twitching dinners.
Left behind, careful as code, false
eyelashes in a dish licked dry.
Belly slice, milky eye plucked
from its socket. What's delicious
is arbitrary, taste of a dim
encounter. What's forgotten:
umbrellas, slim friends
clustered in a bin. Tissue shed
from presents, bobby pins, blister
packs of birth control, headphones,
membranes, hair. The server sleepwalks
with a platter of jewel-toned tuna
and bay scallops. The chef's hollow clap
between pieces, his bowl of water.
A plum sinking in shochu stops
all speaking. A party where nobody comes
is a sad birthday, sparkler unlit,
kettle brimming. We missed
the parade, the concert, the busker,
the meteor shower. We prop it up
and tear it down and prop it up again.

THE HOUSEPAINTERS

They sit in the sky with their brushes,
acrobatics to drench every spot.
They piss in black toilet bowls
and elbow antique doorknobs, text photos
of the resident Saluki to their girlfriends.
They survive clips by trains, sporting
headphones on shortcuts. They fantasize
about cold water and the choreography
of falls from ladders, imagine their poses
in the air. They weigh the cost of middling
injuries. They line up at the free pool
with children who touch their tattoos.
They drink in the park at sunset, bring home
the empty cans. They sleep easily
and deeply, every lamp still burning.

THE HURRICANE SIMULATOR IS UNDER CONSTRUCTION

A child's bike hidden in the bushes. A bus gleams
at the crest, through lattice and methane, forsythia
that has burst and fallen, shaken to the ground by the giant
machine. Craig next door built a pond near the lilacs,
rented a mini backhoe, poured goldfish like punch.
Asks us to appraise his wife's artwork

while she holds her breath in the next
room: mirror, sea glass, glue gun.
Kyla was sunbathing beneath the naked
basketball hoop when the cops got called
on the couple with all the strollers.
Finger cocked to her temple. Chipped lions

eavesdrop on the parrot who repeats
my mispronunciations, preens
his emerald wing. My self-hatred pulls
up a seat, picks at the lace of a fried egg.
Hovering kids are given bubbles,
dented footballs, long tubes of neon

powder. The bicycle pump
takes a break from its wheeze, and a dog's throat
vibrates at bright uniforms, their stop/slow/stop,
the ring of pylons a studded collar.
I remain bound to the rumbling

that surrounds and inhabits me. Levels
subside, leave a slick mudflat pocked
with crab legs, mussels, one sand dollar.
Glass beads gather at the bottom of a pot.
Thunderclouds move in, curls soft
as a Labradoodle's, a rasp at the screen

door, an eardrum rupture. I'm completing
a report about plants, but am stuck at the part
where I'm asked to measure them. I have a ruler

in a drawer, and so much string—but how
to make my brain speak to my arm
so it moves like the yellow crane?

A GOOD CANDIDATE FOR SPACE CAMP

At a leather anniversary, no trace of paper. A tricky address, and maps quit unexpectedly—is tying string around your finger to forget still a thing? I mean, to remember. Belittled in the small print, it was a good night to unravel, to dispatch a set of arrows with your eyes and watch them sink in. Distressed jean jacket meets infinity scarf. A congregation of six, gifts of salt, silk, clocks, moonstone, vows over a firepit. In the range hood's shadow, he spilled about stripping in the forest while waiting for his dog to be fixed, then backpedalled. The crystal ball said, *Nice try*, and for a second summer was over, yellow leaves smacking the patio door, solar eclipse streaming from a laptop. The credenza groaned and we were catapulted into the atmosphere, wondered what we'd come into contact with next.

NOTHING'S WRONG

Three daddy-long-legs skitter in the tub
of a two-and-a-half-star motel. New girlfriend
tests the plastic lounger, sun a pale grapefruit
over the VACANCY sign. All this light makes
a mess, soaks loose screens and splotches
the parking lot like puddles of shed clothes.
Water droplets stare googly-eyed
from sweating glasses. You're facedown
in the grass and I can't tell if you're whimsical
or panicking. I've waited ten years
for what I wanted to hear and just last night
let it go, when a bouquet of baby pheasants
startled it out of me, shooting arrows
from the marsh. Hail fell
in August and I nursed an urgent
hunger, scanned deep-fryer menus
at four a.m. Tell me that whirring
isn't coming from inside, that it's just
the wires and the highway,
passing without pause.

CURRENT MOOD

Wednesday Addams frowning.
Cinderella giving the finger.
A piece of toast that looks like it's frowning.
Serving brunch to a soccer team.
Speaking tenderly to a pickup truck.
A prom dress stuffed under a coffee table.
A cat smoking a cigarette on a pile of money.
A laundry basket filled with french fries.
Ping-pong balls instead of eyes.
A sandpiper wearing rubber boots.
A troop of kangaroos drinking martinis.
A skeleton shrugging, like "Don't ask me!"
A model standing in a field of wheat.
A toddler conked out in his spag bol.
A chain of monkeys linking paws as they rise from their barrel.
A ball of hair suspended from the ceiling.
A staircase to nowhere, lit in heavenly pinks.
A bored baby brother in the adolescent mood disorders clinic.
A beach ball skimming the surface of a puddle.
A rotisserie chicken rolling down Division Street.
A bouquet of selfie sticks on the Grande Canale.
A mood ring stuck on battleship grey.
A seashell cradling a clutch of pills.
One day late according to period app.
Liable to commit libel.
At risk of pyramid schemes.
A fancy dog on a chaise lounge.
2 on a scale of 1 to 10.
Secretly joyful, temporarily mad.

STRONG SUIT

Specks of glass and fool's gold flash
from roadside shrines. I'm driving
into pink: trifle swirl, cherry drop,
swipe of bloody sky. Back at the bachelor
apartment, garments languish in dish soap,
stained elastic, hoping to cleanse
with distance. I'm going to need
a wide berth, or a small boat
unmoored on a tea-coloured lake.
From the sucking silt, rage rises
to perforate the surface, and the fairweather
fly south, chasing what's cozy. This season
is oversaturated, my closest confidante
an old-fashioned clothespin, smiley face
markered on the bulb. I bathe
in sediment, a round stone
balanced on each brow. A fine rain falls
on the gas station—refuelling
has never been my strong suit. The door
slams my thumb, and the pain, which is
nothing, knocks something else loose.

ROLES & RESPONSIBILITIES

1. Smash together day-old cakes, smother result in whipped cream, and label *trifle*.
2. Impersonate teen humanitarian.
3. Sit quietly in empty room. If still alone after one hour, leave.
4. Stuff one thousand envelopes and ward off advances with witchcraft.
5. Cry in broom closet while eating stolen lemon tart.
6. Evict pet ferret from brunch, despite unfinished fruit salad.
7. Alphabetize children before inserting into pool.
8. Count legal tender and log variations from the norm.
9. Rinse ketchup and sour cream from silver ramekins, pink sink.
10. Impose drastic physiological and homeostatic changes.
11. Return lickety-split to previous form.
12. Wear high-contrast uniform to prevent confusion.
13. Read under red strobe light until last call.
14. Plug bread limbs and whiskers into bread body to construct bread mouse.
15. Shade bubbles on fifty-page questionnaire regarding own mental illness.
16. Make a list of places where work has been sought.
17. Safeguard appropriate distance between animals and public.
18. Tighten smock and picture a freckled moon.
19. Punch numbers into register while recalling that porcupine, that nodding raft.

ARCHETYPAL STORY

Mostly too tired to explain
from the beginning. Once upon

a plinth in the sky, dry lips
opened a slit to breathe

fog. Drove past trophy
and engraving shops, overlap

of glow signs, quick
neon speeches that split

our attention. Feral cats
circled the porch, tapped out

emails that said fix this
in all caps. We down-

titrated our need for rebirth,
flew over the bridge

and into the bloated
middle, the digesting snake,

trying to accumulate enough
for an ending, like: they picked

up their weapon and went
on their way, or: The End.

Satin edge of a blanket
rubbed raw to its knitted

bones. I spy an antique
pillbox, crushed opiates, Botox

in the dentist's throne.
Big garbage day, treadmills

kissing oil drums, orchid
wobbling between locked thighs

in shotgun. Telephone poles
a chain of linked crosses.

Marshy hope beyond two posts, mesh
stretched where something got out,

swatch of snagged fabric flapping behind.
Didn't see the water once.

DID WE BLOW A FUSE, LOVE?

    We kept getting caught
in the cord hanging
       from the spent bulb.

    It snowed inside the subway,
blurring newsprint
       and aging dark hair.

    The mountain bristled.
Fractured voices hurtled
       past us as we climbed.

    A cardboard ice cream cone,
weather-beaten, called us
       back home.

    In my dream stood a deserted
church, a rabbit hutch,
       a blueberry stand.

    On moving day, defrosting
refrigerators flooded
       the neighbourhood.

The party fell down
a flight of stairs, but its favours
remained intact.

Smoking at the casse-croûte
counter, half a crossword,
stuck on *trout basket*.

Each booth had its own
jukebox, sad story,
invisible number.

Her eyes were red
as she wrapped cheeses. A building
caved in on TV.

They arrived late with armfuls
of grapefruit, yellow spinning
in every direction.

We placed a classified ad
in person: *dripping vines,
nothing included*.

      I took my broken ankle
out for a crispy spinach sauté
          and a prickly drink.

      He pressed the buzzer,
filled an ashtray,
          forgot the record.

      In my dream
there was a cuckoo clock,
          a love match, a campfire notebook.

      Wind thumped at half-open
cab windows with the ferocity
          of beating wings.

      The aircraft played this trick,
murmured: *You are being carried*
          quickly and easily into another phase.

RITUAL FOR GETTING RID OF

Fishing in my memory's pit to reel in a book I read as a child is trying to grip a splinter that keeps sinking into my thumb I might find it sweet today but its materials were what I plundered to situate my darkest imaginings the characters now dim I am left with mushrooms and coyotes stirring marbled pools in the woods that matched my own woods where far from the egg timer I inhabited moss villages shepherded out at dusk by the trunks' neon birthmarks sleeves caked in the hook-and-loop capture of burrs it was here I learned the limits of my body how long I could think before a hornet whined in my head how many hairs I might pull from my skull before a clearing formed and which other rituals I might enact as well as the limits of my empathy being unmoved by the dead rodents in the woodpile the girl up the street with a hard home life whose voice grated the boy who smelled sour and edged me into a ditch where I played dead for a few moments drawing blades from the earth watchful for white tips when the perfect amount of force was applied scaling the dirt shoulder in time for the fuming bus in the book there was a witch of course the closest I could find in what appeared to be real life the woman next door draped in beads and swinging crosses whose daughter bolted she led me to a litter of kittens squirming in her basement next to slices of pink insulation the posters were a skin on the neighbourhood gradually peeling away the book cast a billowing hammock to enter the story so wholly I must have been comfortable I saw a red light beating from a distance and felt a crisp crown placed on my head I am bumping into chairs in the dark stumbling down a well lit with candles and lined with encyclopedias I am singing the circle of fifths while a thumbnail of a deer pauses against a smoky sky and a tree drops all its berries at once there was a dress shedding feathers in the forest and a child with tiger facepaint and math unfolding as kites or origami I'll never know what they erased but can recover these relics if I unfocus my eyes enough resuscitate them with my mouth and crossed vision

USING A PART OF YOUR BODY, WRITE A LETTER IN THE AIR

I only remember sunsets in elevators.
I see another person
headed to the fifth floor and think we must
have something, possibly everything, in common.
A mystic in maroon scrubs dumps Thursday
into a fluted cup, whispers,
Where is the justice here?
I look around. I rise
to fetch the object I'm holding.
I experience transference
toward the potted plant in the lobby,
its placement by the vending machine
throbbing with purpose and joy.
When it wilts, I sleep for days, haunted
by the tickle of tendrils,
parched for phantom flavours
of chlorophyll and milk.
Every day, a woman mops the floors to gleaming.
I wish she had news to tell me.
A beam bounces the length
of the Scrabble board, letters
stoic in their benches.
I petition the walls for another day.
They lean in on me, greenly.
My little window slides open.
A pudding cup sails by.
In the solarium, I curve
like a blown stem, try random swoops,

finger painting in space.
A voice says, *Hang in there, kittens*,
mouths us back to our spots.
The boy disguised as a flower,
his hair a crown of spikes,
sees roots sprout from the ceiling
in cursive. We flex our slippers,
pompom and fleece,
but rather than spelling
or any clear reference,
it looks like movement;
it looks like a dance.

INTERVIEW BEFORE DEPARTURE

What belongs in the memory palace?
If you kept me a secret, I get you.

How smart is the machine?
Burrow into the heart of the collective.

Is it wise to sleep on an exposed root system?
Trees can and do share information.

Can't you take a joke?
A facecloth in a bucket of bleach.

When a letter is burned, what happens?
The sofa we've always wanted.

Is the voltage enough to kill an ant?
The smallest of entry points will do.

Is a head cold a valid excuse?
There is more than one way to resist.

Has the canoe filled with weeds by now?
I heard rehearsals from underwater.

What has a siren done for you lately?
I am the shipwreck between two beauties.

Does a title equal goodness?
Name it and watch it crumble.

FORCE FIELD

For years none would come near
me, as if I radiated a bubble, a weightless

second skin. So much solo time in hallways
that when it did begin, I was quick to console

after a confession of murdered parents. Turned
out they were teaching math in Scarborough,

but it's a good opening line. I waited
for the fists hailing my door, shut roses,

to falter. I walked unwavering past the stables
on Bell Road, wary of alarming sleeping horses,

as footsteps called my name and skipped
stones against my ankles in the dark.

I considered myself too fearless
for a whistle. I agreed to e-mail only the band

address until I passed the test, which I was told
was a test, one where new girls are riddles

to be solved collectively and discarded.
I allowed my head to be held

down. I did not want to be this person
any more than I wanted to be the person I was

before, doodling on scraps of paper, snug
and dumb, while the rest of the world

seemed engulfed in something blazing and singular—
their own kind of particle barrier. The energy

it takes not to fall silent or explode. How we've
all been fooled forever about where our powers lie.

LITTLE PROOFS

1

A single episode
Glints in the periphery
Mailbox dazzle
Shocking burst of text

2

Glisten close
The cake is in the trunk
If I can't see you
You can't see me

3

We snip the breeze
Into festive shreds
You scale a yellow hill
Pedal hum

4

Just putting out some feelers
Flexible probe
Your face in the dark
Opens, closes

5

I had an inkling
Rubbed a paw
Forced the door
A paper avalanche

6

Earthshine vibrates
A wading pool possessed
Moist conscience
Ragged over a branch

7

Overthink until the lichen
Smiles, lips stained blue
Melting popsicle or frostbite
No glance back to check

8

Here comes the sludge
It lumbers on top of us
Like we're air, pumps
The organ's pedals

9

Tough coconut
You wanted it
The hair and skin
The pleasing slime

10

Twin sensors stroke
The perimeter, agree
On rolling tinfoil
Secretions from a tree

THREE

IF LIGHTNING WAS A THING YOU COULD PLUG INTO

My biography keeps shrinking. Soon it will be only the first letter of my name, a twist of hair on paper. I write a list of thank-yous in advance, people whose benevolence I can sharply envision. I once ate a plant that made me see fur everywhere: on pointed rocks, blanketing the side of a tent, sprouting from all your arms and faces. I am part lightning, I told people, and wasn't joking; I thought pursuing its charge could provide a natural and plausible conclusion. A shepherd chased a flicker into the woods and returned panting and soaked, constellations caught in her eyes. An event with honeysuckle and bunting would prolong our short future, I'm sure, with its toasts, jokes, dropped napkins, duets. Electrostatic brain, electric blue wig, ready to party. Instead, I've begun warding off the elements for the first time since childhood, ponchos and gloves, brims that droop, creams that spray. Patterns repeat, in wallpaper and in the wild, the same shapes forever gusting by. Every time I see my insides on a screen, I think: well, this looks familiar. When a seahorse bobbed into the frame I was hardly surprised. What is me and what is broth, thyroid, fortune, oxytocin, demons, pectin, polyurethane, nori, soil? What gets through to you—hiccups nestled in my hiccups, heartbeat quicker than my own? I hope you can't see the dreams you give me. I lost my breath at the foot of the stairs, saw it whistle out the front door, a storm strobe-lighting the street. I made it through a nine-minute dirge with no rests, inflated the inner tubes of my intestines. Gleaming lung bronchi broke up the sky. I blew up balloons until the room was rubber. Being sent down the river on an air mattress wouldn't be so bad, fronds brushing our faces, snapping at the stems. I banished every extra before I multiplied.

SPELLS

This all happens where houses threaten
to slip into the sea, where no one crosses
on the stairs or sings at the table, where
odd scraps get socked away, a sparrow
in the freezer, nestled next to
the tequila, where the street
sweeper roars by under a strawberry moon.
No matter where, we don't wash
for days and bonfire rises from every layer
of us. Chanting by the kindling, songs morph
into visions, into butcher-paper sketches.
Compatible zodiac signs congregate
around the kicked-in door, now
a slapdash picnic table. Words guard us
against unwelcome thoughts and shifty
visitors. Fragile alchemy gets a baby
to sleep, powered by the noise
machine from the bright chain
store. The power button flickers
blue, and we have entered the right
ventricle. Thirteen beats
before the curtain drops. Our lives shift
softly every time a new one arrives,
high beams interrupting the performance
and stalling the pendulums, the inner workings.
Small fuses, we disconnect and reconnect, willing
the charms we've created to catch.

INTERPRETATION OF SYMBOLS AFTER MONTHS WITHOUT SLEEP

1.

While busy moving all their furniture around, the upstairs tenants were conscripted to join the Natal Day parade. Walking was more difficult than they remembered, the exertion and lack of clear borders between their bodies and the sea smoke. They found a canal to merge with. It was decorated with chatter and baseball caps and orange flags on wooden sticks, and they ended up in a place different than they had expected, much noisier and harder to understand, a funhouse. There was no one familiar, except a young person who relayed with delight she had gotten a nose job. It had been in the way, she explained. Back in the city, the hills were oppressive. Their parents waved from a high-rise, wearing peony corsages.

The parade is a school, the canal an epic poem. The fog, baseball caps, wooden sticks, and peonies are themselves. The flags are shouting, the nose an oppressive hill, the parents stray puppies. Orange could be any colour at all—everything was in black and white. The unfamiliar place was just a day, according to the calendar and the sun, the noise from all the insects who can either be squashed or ignored or shifted onto a sheet of paper, which becomes a ramp to a vaster and leafier kind of place to survive.

2.

In response to an intermittent soft beeping, someone said testily: *I am not a beep technician.* The room I was in charge of erupted. I kept asking the same question and it hung there like wet tights dripping in the shower. I sat at a desk with a lamp attached to it, under a painting of the ocean with its own spotlight. Someone vomited and I had to clean it up with rags, and then I vomited. A jump in time, and a friend and I sat on a padded bench in a gallery, facing an oil nude, catching up on important information.

The answers all exist now; they have been recorded. The problem is that they conflict with one another. The vomit is important information. Anything looks more impressive with a spotlight. Outside the frame of the nude, there was more to see: threaded fish bones, a shaggy chair, woven clouds shaped with wire, lacy floaters, a plastic quilt. Reach for any of these and they disappear. Standing at the window, I saw a smoke detector on the sidewalk, trailing torn wires like jellyfish tentacles. From far far away, a sharp cry pierced the membrane.

PLAY POLICE

play police
listen to your

rubber broccoli
a kid bit, sped to

crash a sunbeam
glass elevator glare

stop takes two beats
to travel a curled tube

we glide toward stars
Science Centre vertigo

if they whirl from view
warnings we do and do

not give our sons
touchscreen lungs

X-rayed bears
my son said

you be the baby
peek-a-boo

with scarves, gloves
a costume bin

from here a lamb
could be a tornado

blood telepathy, palm
trace to take home

April and winter still
April and winter still.

DEAR STRANGER

1.

I painted a message in milk.
I bit it over a candle,
watched it cook.
Two windows were two
eyes, lidded by rolled bamboo.
The house blinked, lullaby,
thrash. Ready to wolf
down clues or consolations,
I looked into a box, a well,
a vault, a Fabergé egg, for an alert,
a jewel, a trapdoor, an escape hatch,
a signal. I lifted all the flaps. I held a glass
to a brick wall, opened my hands, hung
my head out the window. A starling got caught
in my shedding hair, then blew into the hose
where the dryer spews wisps, just
snuffled up his fear
and started nesting.

2.

On all fours on mossy tile,
I entered a cave and then entered
a cave inside that one, and so on.

To figure out what you have left,
emerge with a perfect
half-octopus,
be sad, and
look around.

3.

Evidence I am still here:
dots and dashes through wet grass.
The balcony's squeak, a bell-shaped
shadow. A beating locket. The body's
slow drip. (Also, the wax
cherub in the garden, the unhinged
gate, reflection in a filled bucket.)

Still halfway submerged in a slow-dance dream
with a stingray, enveloped by floppy wings.

4.

Flat, my fever
leaving me, my milk leaving me, gulls
shout into the shipyard's heave,
windchime mobile crowning

a garbage heap. My hipbone pulled
like a drawer handle, opens to reveal a cluster
of pearls, pulsing there.

5.

We made it to the park.
Periwinkles lined up
along the wall, spiral
homes of threads and wrinkles.
A duck's green shine rocked
past the model cruise ship, model
chapel, imagined mini-life.
A cute life, of cutlets and vignettes.
Eyes pursed, snipping chives, spinning a lazy

Susan. O Smokey, you never rest
but you dozed through the hot
air balloon tipped on its side.
The first time you laugh I cry.

I began sorting my spam
and dissolving it in the pond:
Dear Serious Friend / a rare opportunity /
I am in dire / can we discuss privately?
A mermaid in a wheelchair
sent a royal wave from the bandstand.

6.

I walked into the sea with my baby
slung to my chest, salt bracing my bitten
calves, my soreness. Last year quicksand,
this year driftwood, littering the beach
with cryptic order. In the forecast, faded
stork bites, cut lupins tied and drying,

and a headstand, feet rising as if gravity
flipped and simply forgot
to fill us in.

YOU BE A GHOST

I'll hunt for the right
filter, the wash to cast you
as you are, slipping sheer
past the night window.

You be a broom
I'll suck up particles and choke
them out the porthole so they scatter.

You be an insect
I'll dome myself around the lid
of your collapsing shelter.

You be a heart-shaped
pond, I'll be a bog,
preserve the pony erasers and dimes
you shook from your pillowcase.

You wear the headlamp
I'll pierce the river, be the flash
of dog's head paddling its length.

You leap a level
I'll stand under gauze,
pretending not to watch while I do.

You make a circle
I'll swim its border, shading in
the clouds while you orbit.

I LIE ON THE FLOOR IN THE DARK

My kid's breath
has not yet loosened
into liftoff. He sleeps
with a whisk, self-soothes
by strumming its wire loops.
Today we watched a Sea King
warm up near the peninsula's
tip, bleary propeller and steel
waves. Like me, he harbours
obsessions: boxes, lids, sticks,
things he'd follow into the dark
without a lantern or tether.
I'm told I have a double. No one
remembers meeting me the first
time, so, horizontal,
I scribble the math on a leaf.
My evil twin goes shopping
for a wig. People do lie
low on dark floors—I've read
about it in borrowed books.
My attention strays
to an improvised bedtime
tale where our heroine swims
from sleep as from
a rocking tide. Her scales
melt away, she walks
on land, she is converted
to this new way of being.
I don't remember how

her breathing apparatus
adapts or if she ever returns
underwater. I am scarce,
shadow submarined,
and if we could drift
here, parallel forever,
I'd allow it. A great white shark
named Pumpkin disturbed
our waters this summer,
not to mention the
man o' war, floating
terror, who beached,
poison dangling there
in the breeze, oversize
jewel-toned dumpling.
Swimmers captured silent
photos of its dangerous
iridescence, then covered
it in sand, with love.

SEWING MACHINE

First, there was a bare freckled
arm against sparkle-grey night, then just-
snipped hair and cells happy
to regenerate. I detached
into the afternoon's dulcet fuzz

until the gong rang. I burst
or leak, scan the field for the lone
spectator, forget the cold speculum
and crackling paper. Together, we've lost

a few parts: fingertip, appendix, placenta,
balanced a tray of drinks over a threshold,
somehow sidestepped the tripwire

that seems to crop up around my body.
Silver dips in and out of the highlands
like a sewing machine's needle

through green. A stitch,
an ache in a shadow place.

MOM'S NIGHT OUT

We sat in the wrong soft-seater, kept waiting
for things to get funny. A foundling
cuddled the puffballs on their hood strings,
while the screen's muted hues recalled
covert Hitchcocks, watched on low volume
as aunts fanned cold cuts over a plastic tablecloth,
the crows lining the roof's lip no longer neutral.

Two theatres down, clouds crystallized
in dark pleats of velvet curtains.
Performers who'd previously had only
their posture toppled from their turrets,
and the most skilled and suggestive dances
unfurled from the limbs of children.
Moms wore anklets of pins to stay

alert, snacked on clicking oysters
from paper bags. Filthy sparrows,
feathered rockets, wheeled through beams,
worked to unstitch a patch of ceiling.
When a seam gave way to a red

planet, plaster dusted down like talc
or salt. One night soon became many.

AUTOBIOGRAPHY III

A labyrinthine worry expands like a suburb

A child gobbled by tall grass

Pops up, laughing, from the attic window

As if anyone ever collides or colludes

Without fraying the border

As if the give in the weave is the way in

The common cold hurts differently

In the windy heat

The knee sees the mallet coming, the foot

Kicks from a diaphanous hammock

The ladybugs have taken over

The maze is tangled and cool

KINSHIP

A bathtub filled with melted snow, a baby's legs churning

studiously. Dusk highlights our dust, crumbs lodged in seams, fairy

door shimmering a little. The baby's hands smell of watermelon, nutmeg,

raw sliced squash. Once I did things like ride a girl's crossbar at midnight.

I crouched in a clawfoot with a budding anthropologist, mapping the faucet face,

prelude to a walk of shame past a cemetery. But yesterday, first bike ride

in a year, my body mine, safety supplies lacking: lightless, helmet loose.

Animal prints and military bedtimes, pastel shaker eggs versus the warm, silent

chicken kind. Race around the swing set of a neglected school, eucalyptus

rubbed into our chests, water passed back and forth, cold mouths. Churning

in my own underworld, words rotting in my gut, the foghorn cuts the drone

of the shower head. A sweet shadow drapes itself across the stairwell

while all the humans stay sleeping, head to toe.

ACKNOWLEDGEMENTS

Earlier versions of some of these poems appeared in *Minola Review*, *Lemon Hound*, *Public Pool*, *The Rusty Toque*, *NewPoetry*, *The Impressment Gang*, *Matrix*, *illiterature*, *The Week Shall Inherit The Verse*, and *Sunrise with Sea Monsters*. Many thanks to the editors.

Arts Nova Scotia provided much-appreciated financial support during the writing of this book.

I am grateful to live, work, and write in Mi'kma'ki, the ancestral and unceded territory of the Mi'kmaq people. I am grateful to LOVE Nova Scotia and the youth at LOVE for adding so much nourishment and creativity to my day-to-day life.

To those who read drafts, brainstormed, and conspired: Sam Sternberg, Alice Burdick, The Common, Fifteenth Poetry, Stephanie Johns. Some of these poems began in Hoa Nguyen's workshops as responses to work by Lorine Niedecker and H.D.; thank you to Hoa for the sparks. For excellent child care and peace of mind, thank you to Sue and my family.

To everyone at Anvil Press (Brian Kaufman, Karen Green, Cara Lang) for giving this book a home, and Clint Hutzulak for the cover. Thank you to Stuart Ross for his friendship, thoughtful editing, and ongoing belief in my work.

Thank you especially to Lachie and to Clem.

Jaime Forsythe's first collection of poetry, *Sympathy Loophole*, was published by Mansfield Press in 2012. Her poems have appeared in *The Rusty Toque*, *Lemon Hound*, *Minola Review*, *This Magazine*, *NewPoetry*, and more. She holds an MFA in Creative Writing from the University of Guelph, and lives in Halifax, where she works with the youth organization LOVE Nova Scotia.